FREDERICK CARDER:
PORTRAIT OF A GLASSMAKER

Frederick Carder, about 1921.

FREDERICK CARDER:
PORTRAIT OF A GLASSMAKER

Paul V. Gardner

THE CORNING MUSEUM OF GLASS
THE ROCKWELL MUSEUM
CORNING, NEW YORK

Frederick Carder:
Portrait of a Glassmaker

A special exhibition
The Corning Museum of Glass
Corning, New York
April 20-October 20, 1985
Glass from The Corning Museum of Glass,
The Rockwell Museum, and
private collections

Cover: Cluthra group, all made in the late 1920s.
Plate bequest of Gillette Welles.
Inside Covers: Flower and stem detail of Number
32, vase of white cased over Celeste Blue; name
of pattern unknown. Probably 1920s.
Back Cover: Tyrian disk. Name inspired by the
imperial purple fabrics of ancient Tyre. 1916.
Gift of Otto W. Hilbert.

Copyright © 1985
The Corning Museum of Glass
Corning, New York 14831

Printed U.S.A.
Standard Book Number ISBN 0-87290-111-4
Library of Congress Catalog Number 84-07201

Design and Art Direction: Mary Lou Littrell
Typography: Finn Typographic Service
Printing: Village Craftsmen
Photography: Raymond F. Errett and
Nicholas L. Williams

This catalog is funded in part by a generous gift
from Mr. and Mrs. Oliver M. Gibson.

CONTENTS

PREFACE

Few glassmakers could look back with more satisfaction and pride in accomplishment than Frederick Carder. He spent eighty-two of his 100 years rediscovering the secrets of earlier glassmaking as he developed and refined countless decorative techniques and created glorious colors seldom seen before.

The objects included herein are the property of two museums—The Corning Museum of Glass and The Rockwell Museum, both in Corning, New York, where Carder worked the last sixty years of his life. Some of the most important objects in the incomparable collections of Carder's glass at these institutions were especially cherished by Carder himself. Selections from his office, from his home, the pieces which he placed on loan to the Corning Public Library, and even the pieces which were saved before the legendary "smashing" at Steuben in 1933 enrich both of these museums; in addition, Bob Rockwell's own desire and determination to form a comprehensive (complete, if possible) collection of Carder's glass has resulted in the acquisition of many unique objects which are now saved for posterity in The Rockwell Museum named in honor of him and his family. In addition, many other collectors have lent, given, and bequeathed beautiful items made by Carder to these museums and to this exhibition. This catalog gives but a hint of the richness of Carder's legacy to all of us who are interested in glass; its quality has been enhanced by a substantial contribution from Mr. and Mrs. Oliver M. Gibson, Carder collectors.

Today the complete portrait of this remarkable glassmaker can be seen in only one place—The Rockwell Museum. The great collection of Bob Rockwell joined with that of The Corning Museum of Glass forms a dazzling tribute to a glassmaker who tried everything in his search for excellence—and succeeded.

Dwight P. Lanmon
Director

FOREWORD

Fred Carder and I first met at the opening of The Corning Museum of Glass in 1951 – he was an awesome figure of 88. To the 25-year-old director of the brand new museum, he was also an awesome resource for both glass history and glassmaking techniques. Imagine me, walking into such a post with only a bit of booklearning, and being fortunate enough to have a mentor who not only had been in Paris when the Eiffel Tower was being built but had worked with John Northwood, Sr., for Carl Fabergé, and had competed with Gallé, Tiffany, and Lalique! I usually spent Wednesday afternoons with him in his office/studio where he would patiently explain one glassmaking technique after another. The preliminaries were always the same. I would knock on the partly open door and say, "Mr. Carder, are you there?" and he would reply, "Where the 'ell do you think I am?" Then, in answer to my first question about how some decorating effect or other was accomplished, he would glare at me with slightly dilated nostrils, fingers drumming on his desk, and say, "Why in 'ell should I tell you in five minutes what it has taken me a lifetime to find out?" Having thus established the extent of my worth, he would answer at length, showing samples, drawing diagrams, making sure that I understood with a frequent "d'you see?" When I did not, which was often, his pained words became separated, slightly nasal, and a good bit louder: "Stannic – chloride – on – the – *hot* – glass – not cold, damn it." The result of these sessions was a large exhibition at The Corning Museum of Glass in 1952 devoted to his life and work, accompanied by a small biography of which I was the author. He was pleased with both, and our relationship moved from teacher-pupil to friends with common interests in painting, gardening, and art history. But any discussion of glassmaking on any level had me back in the dunce's corner.

In 1973, seventy years after Steuben's founding, I followed Arthur A. Houghton, Jr. as president of Fred Carder's glass company (what invective might stream forth if he had known!). Thus, in addition to mentor, legend, and friend, he became, after the fact, my predecessor. I see him now as the man who insisted on superb craftsmanship at Steuben and on the highest quality of material. Men who worked directly under him have enormous respect for him; what he demanded has become a living part of Steuben. In the pages that follow, Paul Gardner, Carder's assistant from 1929 to 1943, brings to life Carder's strengths and virtuosity as no one else could. It was these qualities which made Steuben the "glassmaker's glassmaker."

In thinking about Fred Carder in the broad context of 20th-century glassmaking, it seems to me that he was remarkable for his absolute and total devotion to one material. The giants among his contemporaries — Fabergé, Gallé, Tiffany, Lalique — all worked in other materials as well. Theirs was a broad aesthetic vision. Jewelry, furniture, whole interiors evolved around their design concepts. Carder, on the other hand, concentrated on glassmaking. He was supreme in this area and, being of a highly competitive nature, tackled any innovation, historic or modern, as a personal challenge. He succeeded so often that he is a unique phenomenon in the history of glass.

Thomas S. Buechner
President
The Corning Museum of Glass

PROLOGUE

Frederick Carder, the founder of Steuben Glass, first became fascinated with glass at age sixteen when he visited John Northwood's studio in 1897 and saw his copy of the famous Portland Vase, considered the world's greatest masterpiece of Roman cameo glass. In later years, Carder loved to recall that when he first saw that vase he was "struck with the possibilities of glass" and "determined, if possible, to get into the business." He "got into the business" the very next year when Stevens & Williams, one of England's foremost glass manufacturers, hired him as a designer at their factory in Brierley Hill, Staffordshire.

So began his glassmaking career—a career which he pursued with vigor and enthusiastic dedication for over eighty of his 100 years. His incredible artistic and technical achievements spanned the decline of Victorian taste, the rise and fall of Art Nouveau and Art Deco, the disruptions of two world wars, and profound changes in economic conditions and life styles, first in his native England and then in America. After half a century as a successful glassmaker, he endured the emotional trauma of seeing his glass go out of fashion in the 1930s, but he lived long enough to see it return to favor in the 1950s and become increasingly popular and coveted by collectors and museums ever since.

Paul V. Gardner

ACKNOWLEDGMENTS

Among my friends and colleagues who have contributed in many ways to this publication and exhibition are Thomas S. Buechner, Dwight P. Lanmon, and Robert F. Rockwell Jr. Present and former staff members of The Rockwell Museum who have assisted me in their specialized areas are Antony Snow, Jensen P. Monroe, Kristin A. Swain, Pamela Michlosky, Mercedes Skidmore, and Elizabeth Anderson. And my special thanks to Robert F. Rockwell III (Bobby) who spent endless hours assembling data from Steuben factory records and worked tirelessly with me in locating and choosing the objects used in the illustrations. I am greatly indebted to another colleague, Thomas Dimitroff, for his help and suggestions concerning the text. I am also most grateful to Raymond F. Errett and Nicholas L. Williams for their superb photography and to John H. Martin, Priscilla B. Price, Norma P. H. Jenkins, and Virginia Wright for their cooperation and assistance in making available the collections and research facilities of The Corning Museum of Glass. I also wish to thank Chloe Zerwick and Charleen Edwards for their editing finesse.

PVG

THE ENGLISH YEARS
1863 · 1903

Born September 18, 1863, in Brockmoor, Kingswinford, Staffordshire, England, Frederick Carder lived and worked within a few miles of his birthplace for his first forty years. Many of the character and personality traits which contributed to Carder's later success developed early, but they did not always work to his advantage at the time. His first experience with formal education ended in disaster. He was expelled from the local Dame school (infant school) for throwing his slate and hitting the principal instead of holding it on his head while standing penitently in a corner. He was next enrolled in the Clark School where David Clark, the headmaster whom Carder always remembered as a "strict disciplinarian," saw to it that young Frederick "toed the mark." In addition to making the necessary adjustments to his precocious pupil's deportment, Headmaster Clark managed to interest him in geography, physics, chemistry, Latin, and Greek as well as "the basics."

But interested as Frederick Carder was in most of these subjects, Clark School activities fell far short of providing enough outlets for his creative energy and artistic talents. Often during school hours he could not resist making sketches of fellow students or the flowers and trees that were visible from the classroom—"especially during mathematics classes." It was at this time that an opportunity to pursue his creative inclinations became available. His grandfather, who owned the Leys Pottery in Brierley Hill, took him to the factory where he encouraged him to "observe the uses of clay in manufacturing processes" and allowed him to "make things" out of this plastic medium. This was just the outlet Carder needed. Soon he could hardly wait for his classes to end so he could hurry to the pottery and "make objects of clay" or "model some of the men in the pottery." Even though his fascination with clay was increasing steadily, he still managed to pursue his studies so successfully that by age fourteen he was "top boy" in his class. He admitted later that "receiving all kinds of honors for my work made me swelled headed, thinking I knew everything."

This conceit, combined with the lure of the pottery, made Carder decide to "quit school and go to work." Although his father objected strenuously, Carder would not change his mind. He was convinced working in the pottery would be "great fun" and begged his father to let him start at once. His father grudgingly "gave in" but, instead of letting Freddie make clay objects and model portraits of workmen, put him to work shoveling coal to fire the kilns. As Carder wrote later, "this was all right for a week, but when it went on for nine months I began to see that I knew nothing." A chastened Freddie then offered to go back to school, but his father reminded him that he had "made his bed and would lie on it." Carder's desire for knowledge

Frederick Carder (R.),
apprentice at Stevens &
Williams, early 1880s.

was now as strong as his wish to work in clay. He soon worked out a compromise with his father, who gave him a better job in the pottery as a reward for his son's agreement to resume his education in night school. Carder kept his part of the bargain and was soon spending three evenings a week studying art in Stourbridge and two evenings a week at the Dudley Mechanics Institute in chemistry, electricity, and metallurgy classes.

All went well for about a year. Carder was eager to learn the pottery business and contributed his own constructive ideas with the encouragement of his grandfather. But Carder's grandfather died suddenly in 1878 and left the pottery to Carder's father and his two uncles. One of his uncles resented Carder's presence in the factory and made his time there unbearable. After months of trying to work out a solution, Carder decided he must leave the pottery as soon as he could find other employment. It was at this critical moment in his life that Carder saw the Portland Vase, became enchanted with the "possibilities of glass," and in 1880, at age seventeen, began his career with Stevens & Williams.

Young Carder plunged into his new-found career with even greater dedication than he had shown at the pottery. As he wrote later, "How well I remember, filled with enthusiasm, I wanted to learn the business in a day and, if that was impossible, in a week...." Forty years later, he admitted he was still learning the business. Even though it took longer to learn the business than he at first expected, Carder's first year was one of great progress. In contrast to the antagonism of his uncle in the pottery, the proprietor of the glassworks "took a great interest in me ... encouraging me at the every turn ... as he desired his factory to turn out the most artistic production of glass in England."

This attitude prompted Carder to initiate several constructive changes. Although the factory was buying the colored glasses used for casings, Carder thought they should make their own. Consequently, he purchased a small furnace that would melt glass in one or two hours. As soon as his furnace was in working order, he began a most ambitious project "to test every known and unknown material that I could get my hands on . . . to that end I had the small furnace going daily, getting sometimes as many as four melts a day. This work was going on at the same time I was designing new goods for the factory." His account continues, "I also made models for any new molds that were required." At that time all molds were made outside the premises since the factory had no machine shop. In order to turn out these molds, Carder bought a small lathe and "commenced to make the models of such things as goblets, etc." He concluded this description of his first year's activities with the delightful understatement, "What with making models, designing [glass], and conducting experiments in making colored glasses, I was kept quite busy."

For the next eighty-two years, Carder "kept quite busy!" In addition to that first year's activities, he spent Saturday afternoons at John Northwood's studio, a routine he had begun at Northwood's invitation shortly after the day he saw the Portland Vase. It had been these Saturday sessions which convinced Northwood of Carder's ability and interest in glass and resulted in his recommending Carder for the position at Stevens & Williams. In 1881 Carder, mindful of Northwood's favors, was instrumental in having his mentor appointed art director at Stevens & Williams. Their combined talents kept the factory in the vanguard of British glassmaking and charted the course which assured its success as a leading producer of artistic glass for many years to come.

Carder's first accepted designs were in the then fashionable cut crystal which he considered "the quintessence of vulgarity." He soon set about replacing them with designs for colored glass. To the surprise of the proprietor, Joseph Silvers Williams-Thomas, these met with almost instant success. From then on, he and Northwood produced a veritable flood of new designs using new decorating techniques—first in the Victorian taste, the perennially popular classical forms, and later in the "new" Art Nouveau styles. Many of these Art Nouveau vases used the cameo technique which Northwood had revived in England with his copy of the Portland Vase, completed in 1876.

Although he continued his strenuous schedule of evening classes and modeling ornamental terra cotta sculptures at "the pottery," on weekends

Annie and Frederick
Carder, late 1880s.

he managed to find time for romance. On May 21, 1887, a wedding
certificate states that Frederick Carder, age 23, bachelor, designer in glass,
married Annie Walker, age 23, spinster, at the Parish Church of St. James,
Dudley, Worcestershire. His bride later reminisced that when they were
"courting," sometimes the only communication she had with her admirer
for days was the rhythmic clatter of a stick being dragged along the iron
picket fence in front of her home as her "Freddie" rode by on his bicycle en
route to work or evening classes. "It was," she recalled, "one of the sweetest,
most exciting sounds I ever heard!"

Being a husband and father (his daughter, Gladys, was born in 1889 and
sons Stanley and Cyril in 1892 and 1893) in no way interfered with Carder's
activities inside and outside the factory. He continued earning certificates
from his evening classes and winning gold and silver medals and other
awards from participation in art competitions. In 1897, he won his Art

Master's Certificate and gold medal for his thirty-inch copy of Hamo Thornycroft's heroic bronze, The Archer. The same year, local authorities in Wordsley invited him to establish an art school there, primarily to give night classes in art and glassmaking to talented glassworkers in the area. Carder was delighted to accept this new responsibility. He had been an art instructor in the Stourbridge Art School for some time and felt that similar classes in Wordsley would be of great benefit to the district's glass industry as well as to the workers. The Wordsley School of Art was an immediate success, and Carder continued as its art master until he left for America in 1903.

Carder's outstanding accomplishments in the artistic and technical fields of glassmaking as well as his contributions to education attracted much attention in the glassmaking industry. In 1902, the South Staffordshire County Council selected him to visit the Continent and report on glassmaking in the German and Austrian glass centers. His published report on this mission proved so valuable to the Stourbridge glass industry that the next year the Council asked him to make a similar fact-finding tour of the American glass industry.

Accordingly, Carder "boarded the OCEANIC S.S. White Star Line at Liverpool on February 25, 1903." His account records that "after a very rough passage...we arrived in New York—in bitter, bright March weather. My first impression of New York was that Doré must have seen it before he designed HADES—with its tall buildings and a devil peering out of every window....What a city!—dirt galore—and smell!—one wanted a clothespin on one's nose...." After spending the night at the "old Astor House downtown," he did some sightseeing and boarded the night train for Pittsburgh. This trip was his introduction to rail travel in America. He was surprised at the "prehistoric methods of the Pullman Company where one had to undress behind a green curtain and sleep in a stuffy box." But he had a productive time in Pittsburgh where he had letters of introduction to "Mellon's boys and...a number of influential men...living, as I found out, hundreds of miles from one another." He was shocked at the "wasteful way" Pittsburghers used natural gas in their iron and steel furnaces and that their streets lights "lit by gas were left burning night and day."

He then went to Washington, D.C., where he "did more sightseeing." Then came another railroad journey from Washington to Corning, New York, "shaking and bumping" through the "dreary brown landscape," climaxed with a pre-dawn, three-hour layover in Elmira—all of which he would "a damn sight rather tell about than do again!" The only bright spot

The Archer. Sculpture modeled by Frederick Carder after the heroic statue by Hamo Thornycroft. Carder won a gold medal and Art Master's Certificate for this work, 1897.

was his congenial seat mate, an "old chap with white hair," who rode with him from Elmira to Corning. The "chap," he found out later, was Mark Twain.

Corning in March was a new experience for an Englishman. But in spite of the soot-laden snow, rutty streets, and icy surroundings he carried out his pre-arranged schedule. First on his agenda was a tour of the Corning Glass Works, which he immediately dubbed "The Smokestack University." He was cordially received by Alanson B. Houghton and other factory officials and duly impressed by their production techniques.

His next appointment was with T. G. Hawkes, president of T. G. Hawkes & Co., a successful glass decorating firm in Corning since 1880. Carder's description of this momentous March meeting was a typical understatement. "During my tour in Corning I visited the Hawkes Co. While in England I had sold them some special blanks. Mr. T. G. Hawkes was anxious to make their own glass instead of buying from others. We had quite a session, and I was persuaded to come back to the U.S. and start a factory in Corning, New York." Carder and Hawkes must have been planning the new factory for some time before Carder's visit to America. The *Corning Evening Leader* of March 11, 1903, announced that articles of incorporation of Steuben Glass Works (named for Steuben county where Corning is located) had been filed with the Secretary of State in Albany, New York, on March 9, 1903. Steuben Glass Works was a stock company with T. G. Hawkes, its president, as the majority stockholder and Carder a minority stockholder.

The four months Carder spent in England between his "session" with Hawkes and his return to "Yankeeland" were packed with preparations for his move and enlivened with varied reactions to his announcement that he was leaving England. There was consternation and a "damned fine dressing down" from Graham Balfour, the official of the South Staffordshire County who had authorized the financing of Carder's trip to the United States; surprise at Stevens & Williams who offered to triple his wages if he would reconsider and stay at the factory; and regrets along with best wishes from the faculty and students of Wordsley School of Art. Carder took all these in stride, stuck to his decision, and sold his house and furnishings, except "for a few pieces" which he shipped to the U.S. along with "my books, drawings, and models—as such were allowed duty free." During all this, Carder was savoring an elation few men are privileged to experience. At the age of forty, he had at last achieved his heart's desire—a glass factory which he could establish and run—the Steuben Glass Works in Corning, New York, U.S.A.

Illustrated address. Presented to Frederick Carder in 1903 by the Wordsley School of Art in appreciation of his "valuable services" and wishing him "every success" in his "new sphere of labor."

To Mr. Frederick Carder,

late Master of the School of Art, and Glass Making Classes.

We, the Committee of the above, take the opportunity of your severing your connection with the School, to place upon record our high appreciation of your valuable services.

As a Student, Assistant, and for the past six years as Head Master, we have watched the growth of your work, and have noted its reflection in that of the Students, and its effect upon the various trades of the district. We have noted with pleasure your skill in Design, which is of a very high order, and which obtained for you the highest possible award of a Gold Medal in the National Competition.

The Class in Glass Manufacture carried on under the Stafford-shire County Council has been most successful, and the practical knowledge gained by the Students cannot fail to have a beneficial effect.

Whilst sincerely regretting your departure, we all most heartily wish you every success in your new sphere of labor.

J. J. Slader M.A.   Chairman.
W. F. Stuart.   Vice Chairman.

Wm. J. Hawkes   James Hill   P. H. Streeton M.D.
Owen Gibbons   Wm. Richardson
E. B. Whitney   B. J. Mason   George J. Carder   Thos. Goodall.
A.H. Richardson   (Science Master) S. H. Fletcher (staff) B. Richardson
                                  Wm. J. Wilkes   Fredk. A. Richardson
                                  Charles Dudley   Secretaries.
                                  W. J. Cashbourd

Wordsley,
July, 1903.

STEUBEN GLASS WORKS
1903 · 1932

From that July day in 1903 when Carder and his family arrived in Corning, he was totally committed to getting the Steuben Glass Works in operation. His night classes, special awards, medals, and over two decades of glassmaking with Stevens & Williams were now prologue. All his talents and boundless energy were centered on his American career. The herculean task ahead was all the more challenging to Carder because of the difficulties it presented. How he surmounted these difficulties is the story of Carder's Steuben glass. It is also the story of Frederick Carder's lasting contribution to the history and development of artistic glassmaking.

The speed with which Carder converted the Payne Foundry building, which Hawkes owned, and adjacent space on Corning's West Erie Avenue (now Denison Parkway) into an operating glass factory delighted the stockholders and amazed interested Corningites. In late October 1903, only three months after Carder's arrival in Corning, the first glass pieces were made at Steuben. These were crystal glass "blanks" for the Hawkes factory to decorate, and they fulfilled the primary purpose Hawkes had for persuading Carder to set up Steuben. Hawkes's business volume required enough blanks to assure the success of the Steuben venture. A less ambitious glassmaker might have been content with this built-in market. But to Carder, simply producing blanks for Hawkes was like creating a canvas for someone else to paint. His creative talents demanded greater expression. As he put it, "I damn well soon got tired of that," and within the year he started producing his own artistic glassware. Whether this additional use of the factory facilities was part of the original agreement or something Carder worked out later, it was apparently sanctioned by the stockholders, and Carder settled into running what he always felt was "his" factory. For all practical purposes, that was what Steuben became. During the next three decades Carder was Steuben's guiding genius, designing the glass and its decorations, devising the batch formulas, hiring and firing employees, supervising all the production, and selling the finished glass.

Steuben factory records show an amazing total of over 7,000 varieties of vases, bowls, goblets, candlesticks, and dozens of other decorative and useful shapes which he designed for sale from 1903 to 1932. In addition, almost 600 designs were made as special orders for wealthy customers and manufacturers, including everything from atomizer bottles to lamp shades. Most of Steuben's output was sold by traveling representatives to leading department stores such as Altman's in New York, Marshall Field's in Chicago, and Gump's in San Francisco. Steuben was also available at the Steuben showroom in Corning. These handblown pieces were produced in more than 140 glowing colors and decorative effects achieved "at the fire" and, after annealing, decorated with hundreds of cut, etched, and engraved designs. The Steuben factory was a success from the start. Carder often boasted that he "bought the materials, built the glass furnace, and retired 40 per cent of the $50,000 indebtedness in the first year of operation."

Aurene, Tyrian, Verre de Soie, Cintra, Cluthra, and Intarsia are only a few of the glamorous names Frederick Carder coined for his artistic glass. All of these and a myriad of other exciting colors, graceful forms, and intricate techniques illustrate the unbelievable diversity of glass this phenomenal glassmaker produced during his career on both sides of the Atlantic.

Every important piece of glass which Carder saw in museums, shops, and exhibitions became a challenge to his artistic and technical abilities. He would sketch the shapes, note the colors, and, as soon as possible, set about trying to equal, improve, or surpass the colors and techniques of each fascinating object. Carder sketched all his life – in sketchbooks, on laundry bills, napkins – whenever and whatever appealed to him, from flowers in a meadow or animals in a zoo to athletic teams in action and fellow Rotarians at lunch. His voluminous sketchbooks, many of which still exist, were not confined to glass. They contained everything from watercolors of models in his art classes to sketches made in museums and salesrooms. Notations on these sketches range from "The Treasury of Priam" on ancient pottery to "Lalique," "Orrefors," and "Argy-Rousseau" on competitors' wares.

It is exciting to discover how he adapted these ideas to his own productions. While it is often easy to trace many of the sources which inspired his creations, as one looks more carefully at Carder's glass it becomes increasingly apparent how often he added his own touch of genius to styles and techniques.

Much of Carder's glass reflects the man himself: forthright and dignified, often elegant – always ready with a positive statement (occasionally relieved with humor). Some pieces show a restraint imposed by a buyer's market, the frustration of every commercial designer. Carder frequently "cussed" this situation, often complaining how he wished he could make only "what he damned pleased." Whenever circumstances allowed, he went a "step beyond the demand" and made glass which started trends and contributed to the development of beautiful handmade glass. The outstanding pieces illustrated show the high standards of beauty and quality for which Carder's glass is so highly esteemed.

Aurene, the first of Carder's now well-known Steuben creations, was inspired by the shimmering iridescence of Roman glass made from the 1st to the 4th centuries A.D. Archeologists tell us these ancient pieces were not iridescent when they were made but acquired their gold and blue sheen from centuries of burial. The name Aurene is of special interest not only because it is one of the most important of Carder's Steuben glasses, but because it demonstrates how much thought Carder gave to coining it. The tie to Roman glass was suggested by the first three letters, "aur" which were taken from the Latin word for gold, *aurum;* then he added "ene," the last three letters of "schene," the Middle English form of sheen. Thus in name as well as appearance Aurene combines its Roman inspiration with the English background of its creator. The name Gold Aurene was patented in 1904 and its companion Blue Aurene (made by adding cobalt oxide to the Gold Aurene batch) appeared the next year. The iridescence of both Aurenes was produced by spraying the glass "at the fire" with a stannous chloride spray sometimes supplemented with iron chloride. Lovers of these glasses have likened the glowing tones of Gold Aurene to the radiance of the dawn and the hues of Blue Aurene to the depths of the midnight sky or the silvery blue reflections of moonlight on a mountain lake.

Carder felt his Aurenes were so beautiful that they needed no additional decorations; only a few decorated Aurene pieces are known. Some early vases made about 1905 have white millefiori flowers with green leaf and vine trailings. Others made a little later have hooked featherings, and an even smaller number have foliate engravings. In the late 1920s, several different vase shapes were ornamented with either leaves and vine trailings or chain motifs in white glass encircling the upper portions.

Carder also liked to use the Aurenes as decorative adjuncts to his other colors. He felt that the richness of Gold Aurene when used as linings and casings or applied in "hooked," "feather," and variations of "leaf and vine" trailings beautifully enhanced a number of colors. One of these was Alabaster, a glass introduced in the early decades of Steuben and named for the translucent white mineral it resembles. Alabaster was often decorated with complete or partial casings of ruby, green, yellow, or brown combined with linings, and trailed peacock feather or other hooked and trailed designs in Gold Aurene. These productions were called Red Aurene, Green Aurene, Yellow Aurene, and Brown Aurene in the factory, and these names, although actually misnomers, are now commonly used to designate these handsome glasses.

Brown Aurene vases and lamp shades made after 1915 quite often had a collar or border of Gold Aurene with a "hooked" decoration of white and brown crisscrossed threads flattened and "rubbed into" the Aurene casing. At that time, Carder called this crisscrossed decoration Intarsia, a name derived from Italian wood inlays called "intarsiatura." Carder's fascination with this name is obvious—he also applied it to two other quite different types of glass. One, which an early catalog calls "New Intarsea [sic] Wares," was made about 1905 and has characteristics reminiscent of decorated Aurene of that period. Only two pieces of this type are known. The other type, which Carder considered his greatest technical achievement, was made in the late 1920s and early 1930s and will be discussed later.

Calcite was another glass named for a mineral and was developed about 1915 primarily for lighting fixtures. These lighting devices were made in a wide variety of shapes and sizes from globular and acorn-shaped shades and lanterns to hemispherical and other bowl shapes. The large bowls, ranging from 10″ to 16″ in diameter, were suspended from the ceiling by rods or chains. The glass in these fixtures was specially fabricated by gathering each bowl in three successive layers before blowing it into its final form. This greatly increased the reflective intensity of the inner surface and at the same time gave a soft, warm light from beneath, which Carder always said was "very easy on the eyes." In addition to shades, Calcite glass was also made with Gold and Blue Aurene linings for vases, bowls, and the very popular flower baskets.

Shades for lighting fixtures and desk and boudoir lamps were an important part of Steuben's business long before the introduction of Calcite. While some early shades were made for gas lights and candle lamps, electricity was the new lighting medium in the early 20th century, and Carder took full advantage of this burgeoning market. The Steuben catalog line drawings show over 600 shapes and sizes of lighting shades ranging from large ceiling bowls and globes to 2″ and 3″ miniature domes and small panels for the chauffeur-driven limousines and elegant town cars of the Roaring Twenties. Most of these shades were sold to lighting fixture manufacturers who incorporated them into their production. A comprehensive collection of these lighting accessories, including individual shades, desk and boudoir lamps, and torchères from some of the foremost lighting companies in the business during the teens and twenties, is found in The Rockwell Museum, Corning, New York.

Another iridescent glass which Carder developed about 1905 and continued to make until the early 1930s was Verre de Soie. This was actually Carder's Crystal glass sprayed "at the fire" with stannous chloride to give it the luminous iridescence which pleased the buying public. Two variations of this glass were introduced through the years, one with pale greenish tint called Aquamarine. When this glass had a Celeste Blue ring applied to the edge, it was named Cyprian.

The success of Steuben's Aurenes and other luxury glass became increasingly evident as the years passed. By 1913, they threatened such competition to Louis Comfort Tiffany, Carder's contemporary, that on November 13th of that year Carder was served a subpoena to answer a "Bill of Complaints" from Tiffany Furnaces which claimed that Carder had used the formula of Tiffany's Favrile glass, which had been given to him by a former Tiffany employee, to produce his Aurene. This was untrue. Indeed, Carder *had* hired a former Tiffany glassblower who had applied to him for a job—but this workman, Edwin Millward, was immediately contacted and signed a long affidavit which stated in part that he "...never taught...Mr. Carder or anyone connected with the Steuben Glass Works how to imitate the methods or processes used by Tiffany Furnaces in making, decorating or coloring glass...while employed by Tiffany Furnaces all I did was to gather glass. No one...tried to get me to leave the Tiffany Furnaces and come to Corning to work for Steuben Glass Works. When I started at Steuben Glass Works that concern was then making the same kind of glassware [as Tiffany]...."

When these refutations of Tiffany's allegations were submitted to Tiffany's lawyers, there was no response for over six weeks. At this point, Carder lost patience with "the whole damn thing" and in January 1914 had his lawyers request the court to order Tiffany Furnaces to answer his interrogations. After a few more weeks of silence Carder received a notice of discontinuance of the action, dated March 5, 1914. When asked about the clash between these glass titans years later, Carder would quip, "they got licked." He usually added that he felt Tiffany himself had little to do with the action, which he surmised was thought up by some "smart-aleck New York lawyers" who thought they could "scare the hell out of those small-town folks in Corning." He also said he and Tiffany met years later at a medal-awarding ceremony in New York City where they "got on famously," shook hands, and agreed to "let bygones be bygones."

The magic of Venetian glass has captivated buyers and influenced glassmakers for centuries. Carder was no exception, and the spell cast by the *maestri di Murano* on his productions was a joy when they were made and is an even greater delight to viewers today. Carder's sketchbooks show many Venetian shapes and decorative motifs such as fruit and flower finials, twisted stems, applied spiral threadings, and horizontal rings in contrasting colors often supplementing gemlike prunts. Decanters, candlesticks, and even tableware exploit the fragility of the medium and the skills of the glassblower to the utmost.

Millefiori, as the Venetian artisans called their revival of the ancient, fused mosaic glass, was another of the intricate techniques Carder mastered. Some of the rare pieces were exhibited at The Metropolitan Museum of Art in the 1920s. From time to time, millefiori pieces attributed to Steuben are offered for sale and, because of their rarity, are a great temptation to collectors. The only positive identification for such pieces is to be sure the fused cane sections are identical to those in similar pieces known to be Steuben.

The name Grotesque, which Carder gave to a type of glass he made in the late 1920s and early 1930s, seems outlandish when compared to his more dignified designations. But when this glass is viewed in comparison to most of his other designs, it really lives up to the dictionary definition of its name and is "fanciful" and "bizarre." This glass is either loved or hated by collectors and viewers. Whichever reaction it evokes, visitors remember it. No doubt this was the reason Carder produced it, and he probably had a characteristic twinkle in his eye when he called it Grotesque!

Acid-etched vases, bowls, and other ornamental pieces were made at Steuben from 1906 to the end of Carder's regime and a few years beyond. Carder loved these creations, some of which were obviously inspired by the English and French cameo glass of the late 19th century, but many more were Carder's own fantasies realized in a wide variety of handsome forms and striking color combinations.

As usual, Carder's choice of pattern names was intriguing. Over 300 names for these etched designs are in the factory records. While some, like Dragon, Dayton, and Hunting, are less imaginative than many of Carder's appellations, they lend a certain aura of faraway places or relate to sporting activities. Others are so downright dull that one can only surmise that there were times when even Carder's creative mind was drained of ideas. His reaction when pressed to name a pattern at such a time might well have been, "Aw, the 'ell with it – call it DUCKS!" Such unromatic names are a source of amusement and conjecture to present-day collectors. Today Fish, Grape, and even Bird Number 1 and Bird Number 2 have their places with the more distinctive names in the history of Carder's etched glass.

In the 1920s, many etched designs were listed as "sculptured." Some of these, especially when made in Alabaster, had raw umber coloring rubbed into the backgrounds to accent the surface pattern and give an illusion of carving. Others listed as "sculptured" seem to be so designated as a sales ploy. Buyers have consistently paid higher prices for wares with elegant names, and the "sculptured" pieces were uniformly more expensive than the wares simply listed as "etched." Carder was a master merchandiser and used many a sales pitch. He often remarked that when a color or shape was not selling as he thought it should, he doubled the price, and "it sold like a shot."

For centuries, glassmakers have obtained many pleasing effects by combining powdered glass, mica, gold and silver leaf, and other materials between gathers (layers) of molten glass. Carder was familiar with these "inclusions" and had used this technique at Stevens & Williams to produce their Moss Agate glass. The Moss Agate made at Steuben was reminiscent of these English prototypes with variegated clouds of colors blended over an inner crackled network, all sheathed in gleaming crystal.

Several other Steuben color effects depended on inclusions of powdered glass. One of the earlier types called Cintra, was produced by rolling the parison (partly inflated gather of hot glass) of crystal over powdered glass of the color or colors desired which had been spread on the marver (the slab on which a glassblower rolls the hot glass). The parison and its layer of powdered glass were then covered with a layer of crystal which not only held the powdered glass in place, but added an optical quality to the finished piece. Cintra was also used as a center for elegant cologne bottles encased in heavy crystal containing controlled bubbles and finished on the outside with prismatic cut designs.

Many other uses of powdered glass were successful in the late teens and twenties. Among them was Cluthra, often described as Cintra with bubbles. Unwanted bubbles have been the bane of glassmakers for centuries. While Carder made fine transparent glass free from bubbles, he also appreciated the use of bubbles to enhance this medium. He often said. "bubbles give life to the glass" and that he "got so sick of people complaining about an occasional tiny bubble which got by inspection" that he decided to put in a lot of bubbles and "make 'em pay more" for the bubbly pieces. The bubbles in Cluthra were due to what he called "a chemical" (probably potassium chloride) mixed with the powdered glass on the marver. This "chemical" produced bubbles in the powdered glass when exposed to the heat of the parison. A layer of crystal glass gathered over the parison, encrusted with the bubbly powdered glass at just the right moment, locked the powdered glass and bubbles forever in delightful suspension.

Other Steuben glasses utilizing "inclusions" were Peach, Rose, and Purple Quartz; Silverina, which has controlled "air trap" bubbles and mica flecks; and Moresque, an ivory glass body covered with broken fragments of paper-thin greenish glass within a thin crystal casing.

Although Carder was as vehement in expressing his dislike for cut glass after he arrived in America as he had been in England, he could not overlook the public demand for this traditionally popular decoration, and he exploited the technique in many appealing ways. The Steuben cutting shop, located on the top floor of the old Payne Foundry building, operated continuously during Carder's management. His skilled cutters produced about 200 decorative patterns on elaborate console sets, stately covered jars, monumental centerpieces, and a few table services. Among the rarest cut pieces are an ornamental peacock, pheasant, eagle, and Pegasus, a fanciful concept of the fabled winged horse.

In contrast to his avowed lack of fondness for cut glass, Carder loved copper-wheel engraving. Most of Steuben's engraved designs were executed by artisans who worked in shops set up in their homes. These men would come to the Steuben factory, pick up the glass blanks and drawings of the patterns Carder had designed, and carry them home. When they had completed their work, they brought the engraved pieces back for Carder's inspection and acceptance. If these pieces did not meet the standards he set, Carder lost no time in "cussing out" the culprits. The experienced engravers, such as Joseph Libisch and Henry Keller, recalled that one such reprimand was usually enough to insure their maintenance of the high quality demanded by "the old man," as many employees referred to Carder "behind his back."

Most of Carder's engraved patterns featured floral and foliate motifs in clusters, festoons, and sprays. Although he used a few favorite buds and blossoms over and over, he bragged that he could "draw dozens of flowers" from memory and varied his designs at will from this floral memory bank. His botanical knowledge was a combination of his art school training and a life-long love of flowers which he enjoyed firsthand in his well-tended garden. As with his etched and cut designs, the pattern names Alpine, La France, Crest, and Strawberry Mansion indicate the diversity of the more than 300 designs listed in salesmen's catalogs and factory records.

Some of the most successful decorations combined cutting and engraving. The V, panel, and other cuttings served to frame and strengthen the more delicate engraved designs. These cut and engraved patterns were done in both matte and polished (rock crystal) finishes. While these decorations are effective on crystal and monochrome glasses, the finest productions are cut and engraved through casings of ruby, green, blue, amethyst, and black over crystal, alabaster, and other matrix colors. When these combinations of cut and engraved designs were used on tall, stemmed goblets and other elegant tableware, they raised the standard of beauty to a height usually thought to be reserved for strictly ornamental objects.

Stemware and other table accessories including candlesticks, compotes, and centerpieces were a very important part of Steuben's production throughout the entire Carder period. Over 260 different stemware designs are shown in the line drawings of goblets which indicated the style for fifteen or more matching glasses. A typical stemware service might include the following items:

Goblet	*Whiskey tumbler*
Saucer champagne	*Appolinaris tumbler*
Claret	*Brandy and soda tumbler*
Wines (red and white)	*Lemonade*
Sherry	*Hollow-stem champagne*
Cordial	*Cocktail*
Half-pint tumbler	*Finger bowl and plate*
Champagne tumbler	*Parfait*

These stemware sets were another merchandising scheme. If a hostess wanted to be sure she was using exactly the right glass to serve her guests, she needed to purchase fifteen or sixteen types of stemware by the dozen, plus the matching or harmonizing candlesticks, compotes, and other accessories.

Carder ran "his" Steuben for fifteen years with very little interference from the stockholders or anyone else. The pattern of success set in the first years continued as Carder designed and merchandised artistic handblown glass to ever-expanding markets. New colors and innovative techniques followed each other in rapid succession. The original ten-pot furnace installed in 1903 continued to operate and was supplemented in 1908 with a sixteen-pot furnace in the enlarged blowing room. The rest of the factory was also enlarged; by 1912 it occupied most of the block on West Erie Avenue from the original Payne Foundry building to Chestnut Street. An expanded work force which included the blowing room, cutting shop, etching rooms, packing and shipping facilities, and office staff now numbered about 270.

The outbreak of World War I put an end to Steuben as an independent factory. Late in 1917, the U. S. government declared Steuben a nonessential industry. This meant the factory could no longer purchase certain raw materials necessary for glassmaking as they had become restricted to the war effort. Carder and the other stockholders were faced with the choice of shutting down Steuben, throwing about 270 employees out of work, or selling the factory to the Corning Glass Works which had long wanted to own this prestigious establishment. They chose to sell, and in January 1918, the Steuben Glass Works became the Steuben Division of Corning Glass Works, which could not produce art glass during the war years.

Carder had not yet adjusted to the sale of "his" Steuben to "The Smokestack University" when he suffered a much greater tragedy. His 25-year-old son Cyril, a lieutenant in the U. S. Army, was killed in action in France on July 18, 1918.

During the next year, the grief of losing his son and the changes at the factory combined to make Carder feel he must get away from Corning and give serious consideration to his future. As the months passed, this feeling intensified, and in the spring of 1920 he and Annie sailed for Europe with the thought they might possibly return to England permanently. After visiting Cyril's grave in France and spending some time in Italy, they went on to England. They found an England greatly changed by the war. After a few weeks with family and friends, he and Annie realized they no longer felt at home there and returned to Corning for good.

Back in Corning, Carder was reinstated as manager of the Steuben Division and within a few months the factory assumed much the same routine as before the war. Carder in his sixties was as active both mentally and physically as ever; he continued to produce new designs and colors in complicated techniques at the same incredible pace he had maintained during Steuben's earlier years. The Gold and Blue Aurenes, Verre de Soie, and most of the other popular prewar colors continued in production and were supplemented with many new delights. Among the new colors introduced in the 1920s were Florentia, Alexandrite, Oriental Jade and Oriental Poppy, Wisteria, Ivrene, and a name he had used before, Intarsia.

Carder considered this latest Intarsia his greatest achievement in artistic glassmaking. In 1916 designers and glassmakers in Sweden had developed a technique called "Graal" in which colored reliefs were cased in crystal and the surface then smoothed. Carder had experimented with this technique in 1916 or 1917, but his commercial production did not begin until the 1920s. His rare bowls, vases, and goblets usually have a floral or foliate design in a thin layer of colored glass enclosed between two thin layers of crystal glass. Only one exception to the colorless crystal layers is known. This extremely rare Intarsia vase has an amethyst design enclosed between two layers of French Blue. Nearly all Intarsia vases and bowls have the facsimile signature "Fred'k Carder" engraved on the side of the bowl or vase. Goblets are often found without this signature.

But as always, tastes were changing. Many European decorative wares in the Art Moderne, futuristic, functional, and other simple styles mostly devoid of ornament, were being touted as "pure art" forms. For years Carder had kept abreast or been ahead of the "new" trends, but these unadorned forms were not to his liking. He felt many were inspired by Bauhaus architecture which he hated and always said was "cut off with a G.D."! Everyone who knew Carder and his often colorful language needed no explanation as to what the letters "G.D." stood for. Another Carder comment was that these designs were made of "straight lines and pot hooks" which in his estimation did not require any artistic skill to put together.

In spite of his earlier antipathy for these styles, Carder relented somewhat after he visited the Exposition des Arts Décoratifs et Industriels in Paris in 1925 as a member of the commission selected by Herbert Hoover, then Secretary of Commerce, to represent the United States. Soon after his return a number of handsome vases and decorative pieces in Art Deco styles appeared in Steuben's production. These styles continued to supplement the new designs offered to the public in the late 1920s and early 1930s.

However, Art Deco pieces and other innovations such as the Prong and Grotesque vases were a minor part of Steuben's production. The majority of Carder's designs were still in the more traditional classical, Venetian, oriental, and English styles which had been the backbone of his business for more than fifty years.

Regardless of Carder's efforts and token concessions to the "new" trends, sales of Steuben glass declined in the late 1920s. The 1929 stock market crash and the Great Depression of the 1930s, combined with a number of other factors which had threatened Carder's status for some time, finally brought an end to his control of Steuben. In February 1932, at age sixty-nine, Carder was relieved of his duties as manager of the Steuben Division and, as he said, "kicked upstairs" with the title of Art Director of Corning Glass Works.

In 1932, Arthur A. Houghton, Jr., great-grandson of the founder of Corning Glass Works, became president of Steuben. He introduced a new type of optical-quality lead glass recently developed by Corning Glass Works scientists. Objects made from the new glass had unusual clarity and brilliance. New designers were hired, and within a few years the name Steuben was associated with the clear, highly reflective lead crystal which is still produced in Corning.

THE STUDIO PERIOD
1932·1963

Having to relinquish command of the Steuben Division while he still felt fully capable of running "his" factory was a major emotional crisis for Carder. His stoical exterior kept most of the public unaware of his real feelings, and the special efforts of his family and friends served to sustain him through his ordeal. Actually, being made Art Director of Corning Glass Works — with a spacious office, a full-time assistant, (author; Paul V. Gardner, 1930 graduate of Alfred University, B.S. Ceramic Art), a secretary, and no reduction in pay — gave Carder far more than most employees, especially those nearing three score and ten, were offered. Only those who have been forced to relinquish power can fully appreciate Carder's situation. Although he was rebellious, he was realistic and grudgingly accepted the inevitable.

Always a man of action, within days he began converting the 20' by 40' office space allotted to him into a place where he could produce not only designs but glass objects. He shoved his rolltop desk and drafting table to one end of the room and his assistant's rolltop desk and smaller drafting table to the other. The space between gradually became a working studio complete with facilities for modeling in clay, casting plaster of Paris molds and, to the surprise of Corning Glass Works officials, a small homemade electric furnace capable of melting glass. The studio in which Carder the artist not only designed the glass but built the furnace and produced the glass was an American harbinger of the so-called Studio Glass Movement initiated in the 1960s by Dominick Labino and Harvey Littleton. All this feverish activity not only served to divert his mind from the unpleasant realities of his new position, but also laid the groundwork for his new glassmaking activity — casting glass by the *cire perdue* (lost wax) process.

The *cire perdue* process had been used since ancient times for metal casting but never to a great extent for glass. Carder had been intrigued before and after coming to America by the *pâte de verre* and *cire perdue* castings of the French glassmakers Henri Cros, Georges Despret, François Emile Décorchement, and others in the late 19th and early 20th centuries. Now for the first time he could devote almost full time to "having a go" at this fascinating technique. With the same enthusiasm and almost the same energy he had exhibited fifty years before at Stevens & Williams and thirty years before when starting Steuben, he vowed to conquer this new technique.

The major problem in casting glass by the lost wax process was the mold material. Carder envisioned a ceramic mixture which would be strong enough to withstand the weight and movement of the melting glass during the casting period and friable enough to be broken away from the fragile glass casting after annealing. This proved to be a monumental task. There was no published formula for such a mold material, so Carder decided on a trial and error program. At times, it seemed he would never achieve the desired results. After months of persistent effort he finally succeeded in developing a ceramic mold formula which filled his needs. Some said he cussed his way through. Actually it was his practical knowledge of materials aided by his stubborn determination which made him succeed. When asked how he managed to continue the project after repeated failures he replied, "Never say die – say DAMN!"

Almost a year and a half after Carder had become the "Art Director," he was making considerable progress in his *cire perdue* experiments and his time was pleasantly occupied with designing for various Corning Glass Works departments, consulting with officials, and bossing his assistant. All these activities appeared to have made Carder's transition easier than at first expected, and his friends and colleagues began to feel the worst was over. Then one October morning in 1933 word spread through Corning and reached Carder's studio that the new Steuben management had smashed all the glass remaining from the accumulated stock of Carder's Steuben. Carder, along with most of Corning, was stunned and unbelieving. But before nightfall, the rumors were confirmed. Sensational accounts in the press stated that the "planned vandalism" had destroyed 20,000 pieces of glass valued at $1,000,000.

In some ways the news of this deliberate destruction was more traumatic for Carder than his removal from Steuben. As he walked to the Glass Works the next morning, he gave little outward indication of his emotional state. He nodded to passersby with unsmiling civility, stomped into his office and proceeded to his desk with only a perfunctory greeting to his assistant. Within minutes after his arrival, he set about expressing his suffering as artists have for centuries, by producing a work symbolizing his innermost feelings. He donned his smock, extracted several pounds of plastic clay from its bin, and started work. For the next two hours, he worked in total silence except for the slapping of his hands on the clay. Then it was done – a life-sized head of Christ, crowned with thorns – and the silence was broken with his almost inaudible comment, "This is crucifix-ion."

Carder might have been spared much of his grief if the true facts of "the smashing" had been revealed to him at the time. The participants admitted much later that it was in reality a publicity stunt greatly exaggerated by the press. Most of the stock which had accumulated during Carder's management had become discontinued items when the management decided to concentrate on the production of crystal and phase out the colored glass. As was customary in most glass factories, many of these discontinued articles had been sold to employees and the public at greatly reduced prices. After some months, the space occupied by these remaining pieces was needed for other uses, and the problem of their disposal was the genesis of the idea that smashing these well-known quality glasswares would attract wide attention and give a lot of free advertising to Steuben's new direction. Years later it was also admitted that, for several days before the smashing, some of the officials had gone through the stock and picked out the most valuable pieces which were put in storage. About forty years later, most of these pieces were taken out of storage and given to The Corning Museum of Glass. Most of the pieces smashed on that October day in 1933 were chipped and cracked vases, crooked candlesticks, and other rather unimportant items which at that time were "out of style" and had not sold even at the bargain prices.

It is possible some informed insider may have leaked the incident's true facts to Carder. At any rate, his anger soon vanished. In a short time, the only visible evidence that "the smashing" had occurred was the plaster model of the head of Christ, which he later cast in glass.

For the next three decades, the perfection of the *cire perdue* glass casting process was one of Carder's chief concerns. Starting with relief panels in *pâte de verre,* he gradually perfected the technique until he was able to cast the unbelievably complicated Diatreta pieces in the 1950s, when he was in his nineties. Carder took the name and the idea from the fourth-century Roman "cage cups," called Diatreta. These "cage cups" were made by grinding away the glass separating the outside "cage" from the inner matrix, a process entirely different from Carder's lost wax Diatreta castings. If Carder had made no other glass than his Diatreta and other lost wax castings, he would have an assured place among the giants in the history of glassmaking.

In addition to perfecting the *cire perdue* casting process, Carder and his assistant were kept busy designing for many areas of Corning Glass Works, from the home plant in Corning to the MacBeth-Evans Division in Charleroi, Pennsylvania. During the 1930s, architectural glass, which Carder had long felt was unexploited by architects and builders, began to be ordered from Corning. Most of these glass units were blown or pressed in iron molds. Carder and his assistant carved the plaster models for these elements, usually to the architect's designs. These panels and other decorative

Installation of lighting panel in the *Daily News* Building, 1930s.

Lighting panel. Intaglio design in Crystal glass. Several of these panels were installed in the elevators of the *Daily News* Building, New York. The plaster model was carved to the architect's (Raymond Hood) design by Paul Gardner, 1930s.

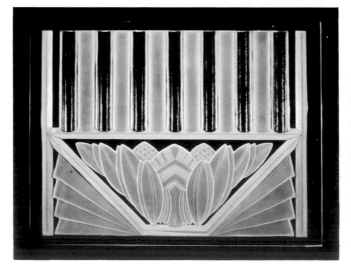

Lighting panel. Intaglio design in Crystal glass. One unit of a repeat border installed below the ceiling in the elevator lobbies of the Empire State Building, New York, 1931. Every other panel was inverted to create a rhythmic sequence. The plaster model for the iron mold was carved to the architect's design by Paul Gardner, Assistant to Frederick Carder.

Full-scale plaster of Paris model of panel by sculptor Lee Lawrie for center section of entrance loggia of the RCA Building, Rockefeller Center, New York, 1933. The model was cut into sections from which iron molds were made.

Complete entrance panel, RCA Building, Rockefeller Center, New York.

pieces were used in many new buildings including the Empire State, Daily News, and Radio City Music Hall in Rockefeller Center, New York City, as well as in other American cities. The most important installation was the 55' by 15' panel in the entrance loggia of the RCA Building, also in Rockefeller Center, completed in 1933. This was made of Pyrex® cast in sections in iron molds modeled to sculptor Lee Lawrie's design. Other projects included redesigning the original Pyrex® ovenware and making the shapes for the first Top-of-Stove ware when the new line was introduced.

The largest lost wax casting Carder ever made was an Indian head 46" high, complete with feather headdress, cast in Pyrex® glass and weighing 800 pounds. This monumental head was cast in the mid 1930s to show Carl Milles, the Swedish sculptor, and officials from St. Paul, Minnesota, that Carder and the Corning Glass Works could cast a glass statue 35' high in nine sections, each about 4' high. Milles had been commissioned by officials in St. Paul to create a decorative sculpture for their new City Hall. Milles's design for this sculpture was to be a figure of the Great Spirit rising from a flaming campfire to the amazement of a circle of Indian braves seated around it. Milles had heard of Carder's success in casting glass by the lost wax process and felt that glass would be the ideal medium for his sculpture. He was particularly intrigued by the possibilities of illuminating the glass sculpture from within and underneath. This was especially suitable to the campfire motif and would impart an aura of mysticism and drama to the glowing figures, especially since their setting was to be in a huge black marble concourse.

Milles arranged to visit Carder in his Corning studio to see if Carder would be interested in the project and brought with him a small scale model of his sculpture. Carder was captivated by Milles's model and the idea of making the world's largest cast glass sculpture. Before their meeting was over, he had assured Milles he could cast the statue. He also offered to cast a glass Indian head about the size of Milles's Great Spirit to demonstrate it could be done. The elated Milles returned to Minnesota and obtained almost instant permission for Carder to go ahead with the trial casting. Carder lost no time in modeling his version of an Indian chief, and within a few weeks the Pyrex® glass casting was resting securely in its ceramic mold where it was to stay during the weeks required for annealing (slow cooling) the 800-pound mass of Pyrex® glass. All would have been well except that the committee from St. Paul came to Corning to inspect the casting several weeks before the annealing period had been completed. In spite of Carder's vehement objections, they had insisted on the earlier date. "It will crack!" said Carder as he reluctantly removed the protective mold and insulation— and crack it did—horizontally through the middle of the head just above the tip of the nose, where the glass is about 8"-10" thick. The committee was duly apologetic for the damage their early arrival had caused and promised to give a favorable report when they returned to St. Paul. They kept their word, but economic and political pressures overruled their recommendations with the argument that the money the project would cost should be spent in Minnesota and not given to a New York State firm. So the Great Spirit and his braves that now adorn the St. Paul City Hall were carved from white Mexican onyx. And the giant Indian head, repaired and suitably illuminated, is handsomely displayed in the promenade gallery on the first floor of The Rockwell Museum in Corning, New York.

The 1940s proved to be another time for testing Carder's "metal." His assistant enlisted in the Navy in the fall of 1942 and was called to active duty in January 1943. Later that year, Carder suffered the loss of his beloved wife and helpmate of fifty-six years. Carder's solace during these trying times was, as before—work.

In 1937 at age 74, he had written, "In looking back over a busy life it seems strange that I should be writing my experiences at a time when the average man lays down his work — I am in good health and feel just the same as I can ever remember. I feel like work, for that and that alone is the one thing that makes a man feel and keep young — that is if he gets pleasure out of his work...."

And he continued to work — casting Diatreta and other glass pieces until 1959 when he decided, at age 96, it was "time to retire." But even then he did not quit working. He reactivated the third floor studio in his Pine Street residence and turned to painting in watercolors and oils. He turned these out at an amazing rate, often inviting friends to "pick out a painting" from dozens stacked on the floor or leaning against the wall. Between times he held court, regaling visitors with anecdotes concerning Lalique, Fabergé, and other contemporaries of his time in glass and in signing Steuben pieces for collectors and dealers. He was greatly pleased that his glass had come back into favor, although he was constantly amazed at the prices "his" Steuben was bringing, even in the 1950s.

Carder continued in remarkably good health and celebrated his 100th birthday on September 18, 1963. As was his practice for fifteen years, he gave a gala black-tie dinner for about forty of his friends which he hosted (so he could ask whom he "damn well pleased") in the top floor dining room of the Baron Steuben Hotel in Corning. On December 10, 1963, after he had spent a pleasant evening in his Pine Street home talking about glass with his friend Robert Rockwell, he climbed the winding stairs to his second floor bedroom. He refused any assistance from his housekeeper with, "I've climbed these stairs for sixty years, and I'll climb them tonight." That night he died peacefully in his sleep.

One of Carder's contemporaries, George Bernard Shaw, wrote "The greatest artist is he who goes a step beyond the demand and, by supplying works of a higher beauty and a higher interest than have yet been perceived, succeeds, after a brief struggle with its strangeness, in adding this fresh extension of sense to the heritage of the race...."

Frederick Carder went a step beyond.

Owners of objects shown in the catalog are acknowledged in the Object List on page 119.

THE
ENGLISH
YEARS
1863·1903

1 Cameo vases. Designed
   by Frederick Carder, produced at Stevens & Williams in the 1880s.

40

2

2 Vase, shaded amber cased over white with applied decorations in "mat-su-no-ke" style. Designed by Frederick Carder, produced at Stevens & Williams, 1880s.

3 Spiral air-trap vases in satin glass called Pompeiian; by Frederick Carder and John Northwood, produced at Stevens & Williams, 1880s.

5

4 Immortality of the Arts, cameo plaque (unfinished) designed and carved by Frederick Carder, 1891.

5 Cut Intaglio vase in Art Nouveau style, designed by Frederick Carder, produced at Stevens & Williams, 1890s.

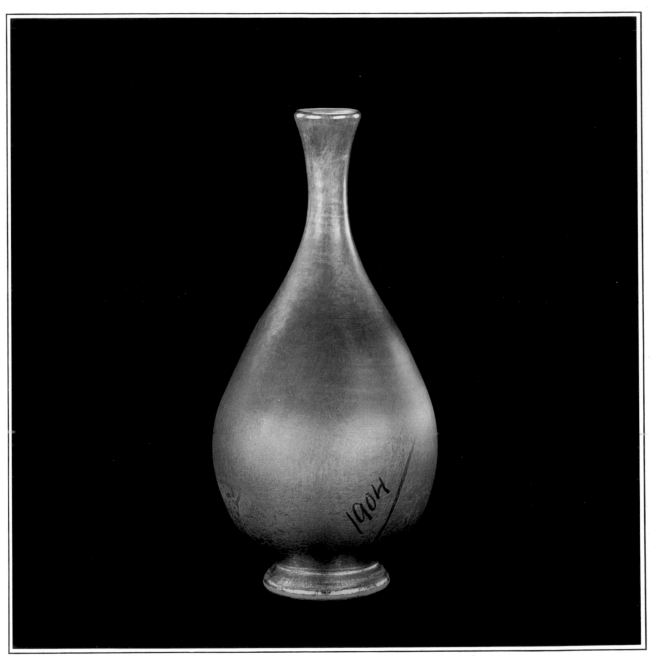

6

6 Gold Aurene vase. The first production
   piece made in this color. The date "1904"
   was marked on the side in wax pencil by
   Carder, who kept this vase in his private
   collection until his death in 1963.

7 Gold Aurene vases. Typical pieces showing
   the range of fully developed color; produced from about 1904 to the early 1930s.

STEUBEN
GLASS WORKS
1903 · 1932

8

8 Blue Aurene group. Made from 1905 to the early 1930s, this handsome iridescent color ranged from a silvery sheen to a peacock-feather blue. With its companion Gold Aurene, Carder firmly established Steuben in the luxury glass market. Most Aurene pieces were marked with the word "Aurene" and the catalog number engraved on the bottom.

9 Decorated Gold Aurene with millefiori flowers, and green leaf and vine trailings. The silvery gold sheen is typical of many early pieces, 1905-1910.

11

10 Red Aurene, Green Aurene, Brown Aurene vases, about 1910-1915.

11 Green Aurene and Brown Aurene vases. Applied millefiori and trailed leaf and vine decoration on both, about 1910-1915.

12

12   Brown Aurene vase. The brown and white crisscross decoration on the neck was called Intarsia when these vases were made, about 1910-1915. It is quite different from the pieces, also called Intarsia, made from about 1929-1931.

13   Gold Aurene vase with Black/Alabaster neck with acid-etched pattern. Very rare, probably 1910-1920.

14

14 Vases, baskets, and other decorative accessories of Calcite with Gold and Blue Aurene linings, about 1915-1920s.

15 Boudoir lamp. Green Aurene, about 1916.

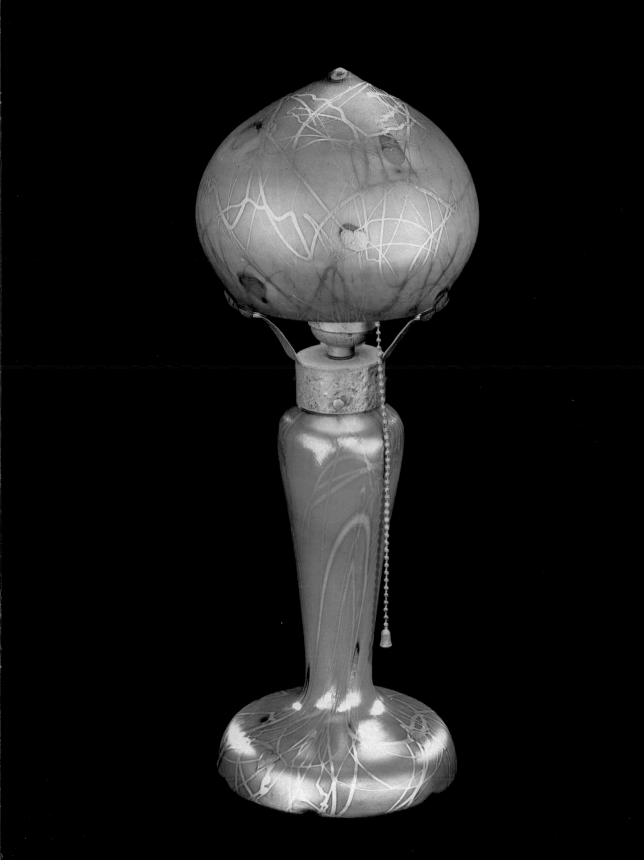

16

16 Lamp, Alabaster with pink, blue, and Au-
    rene feather decorations, 1910-1915.

17 Verre de Soie group. Carder introduced
    this iridescent glass in Steuben's early
    years. Its silky sheen had instant appeal,
    and it became one of his most popular
    productions. About 1905-1930s.

18

18 Gold Aurene vase with tooled decorations and swirled ribbing. Possibly unique, about 1910.

19 Candlestick of Gold Aurene. Art Nouveau style, 1920s.

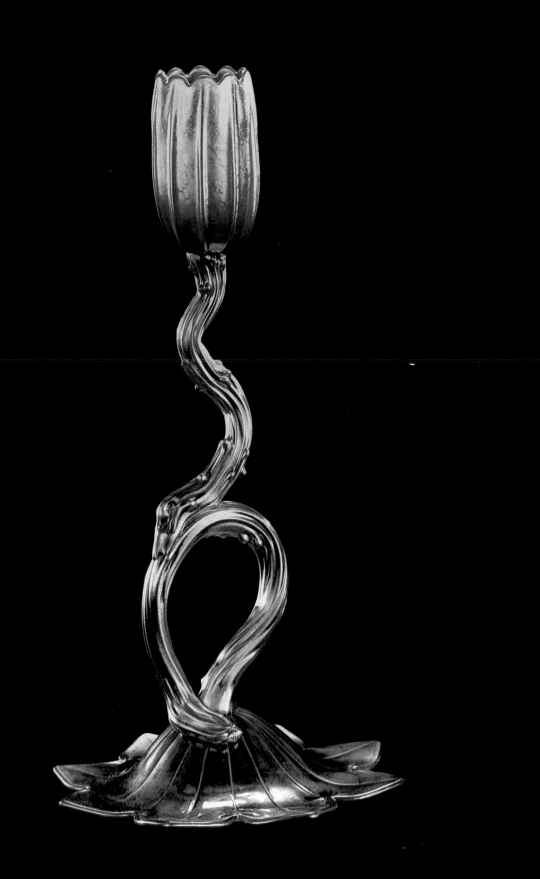

21

20 Decorative accessories and tablewares. Topaz with rings, prunts, and other Flemish Blue decorations in Venetian style, 1920s.

21 Vase/Water-lamp base, French Blue. The metal fittings holding the light bulbs and shade were designed for easy removal when this object was used as a lamp base and filled with water. The water greatly enhanced the optical effect, perhaps suggested by the colored glass ornaments in pharmacists' windows in the early twentieth century. 1920s.

22

22 Console set. Light Jade Blue and Alabaster, Venetian style, 1920s.

23 Millefiori group. Very rare, about 1915-1930.

24 Drinking glasses. Hand-blown in colors including Verre de Soie, Rosaline and Alabaster, Oriental Poppy, and Oriental Jade. 1920s.

25

25 Tyrian disk. The purple shading was developed by reheating that portion of the object during the forming process. The imperial purple fabrics of ancient Tyre inspired Carder to coin the name in 1916 when most of these rare pieces were produced.

26 Grotesque glass group. Amethyst, Flemish Blue, and Ruby shaded to Crystal. The two vases in the center are the tallest Grotesque vases known; late 1920s-early 1930s.

28

27 Grotesque bowl. Dark Jade Blue; extremely rare in this color. About 1930.

28 Grotesque bowl, Blue Aurene. Rare, about 1930.

30

29 Hexagonal vase. Black cased over Celeste Blue, acid-etched in Hare Bell pattern. The background is also etched in a random dot design called "double etched" at the factory. 1920s.

30 Vase of Plum Jade. A rare, double cased color made by combining Amethyst cased over Alabaster over Amethyst. Acid-etched Peking pattern, 1920s.

31

31 Bowl of Black cased over white Cintra
    with faint greenish tint. Mold-blown,
    acid-etched in Supino pattern, 1920s.

32 Vase of white cased over Celeste Blue.
    Acid-etched; name of pattern unknown.
    Probably 1920s.

33

33 Moss Agate vase. Probably 1920s.

34 Cintra group. The striped vases are a variation of the Cintra monochromes; Carder sometimes called the striped vases Orverre. All made about 1917.

35

35　Cologne bottles. Cintra center surrounded by colored threadings and controlled bubbles; cased in heavy Crystal with massive cutting, 1920s.

36　Cluthra vase. Shaded green to white; applied crystal handles, late 1920s.

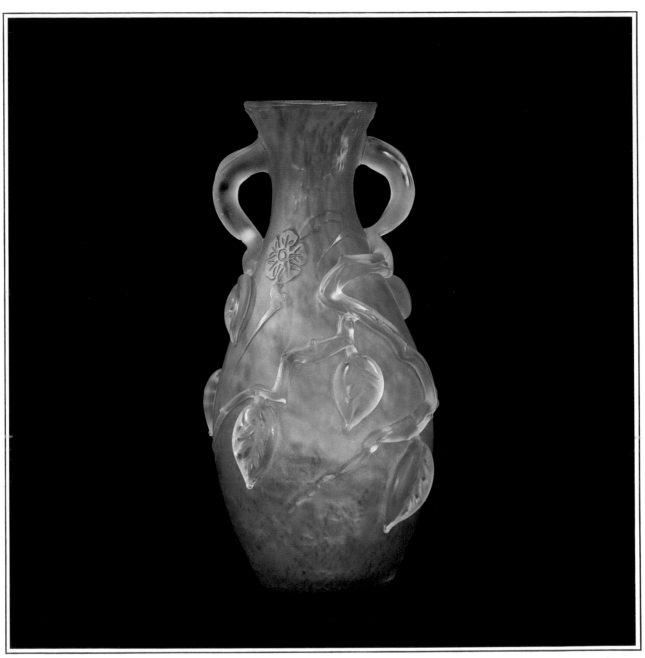

37

37 Rose Quartz vase. Reminiscent of Cintra and Moss Agate, the Quartz glasses (made in several colors) combined an inner layer of crackled Crystal covered by finely powdered Ruby, both enclosed in a Crystal casing ornamented with tooled Crystal leaves and stems. After annealing, an acid-etched decoration was added, and the entire exterior was given a satin finish with buffed highlights. Late 1920s.

38 Silverina bowl and candlesticks. This glass featured overall air-trap patterns and mica flecks. Late 1920s.

40

39 Covered jar of Moonlight glass. Cut deco-
    rations, late 1920s.

40 Place setting, Light Amethyst cased over
    Crystal. Cut Wheat pattern, 1920s.

41

41 Peacock. Crystal with cut decoration,
    about 1930.

42 Vase, Flemish Blue cased over Crystal. Cut
    decoration, 1920s.

43

43 Console set. Flemish Blue cased over Rosa. Pieces in this color, called Alexandrite by Carder, always have cut decorations usually with panel, flute, and diamond motifs; 1920s.

44 Console set and vase. Rose cased over Alabaster; engraved York pattern, 1920s.

45

45 Goblet. Medallions of Flemish Blue over Crystal. Engraved Music pattern, late 1920s.

46 Stemware. Cut and engraved decorations on Crystal and on Ruby and Black cased over crystal (l.-r.) Fisher, Chateau or Melba (?), Cordova (?), Alpine patterns; Crystal glasses (right) engraved in Strawberry Mansion design; 1920s.

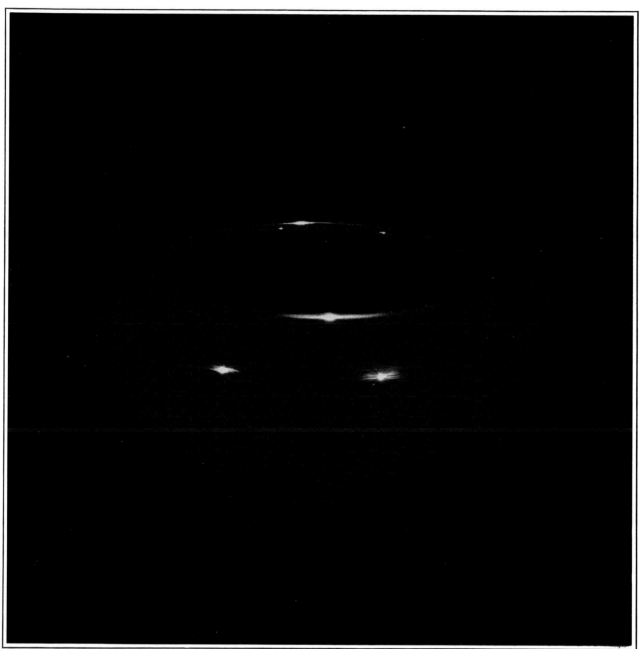

48

47 Mandarin Yellow group. Inspired by Chinese porcelain. Only a few pieces were made because the glass was unstable and cracked or broke usually within a few days or months after being made; probably about 1916.

48 Rouge Flambé bowl. Very rare color inspired by Chinese *Sang de Boeuf* porcelain; first produced in 1916 and revived in 1926.

49

49　Oriental Jade (left) and Oriental Poppy
　　(three pieces on right). Late 1920s.

50　Florentia group. The leaf-shaped elements
　　were prepared from powdered glass in the
　　colors shown, usually green or pink; these
　　were picked up on the hot Crystal glass
　　gather (parison) and worked into finished
　　form. After annealing, the surface was
　　given a matte finish by acid etching or
　　sandblasting. Late 1920s.

51

51 Candlesticks and compote. Ivory with
    Black bases, about 1930.

52 Lily vase. Ivrene, late 1920s.

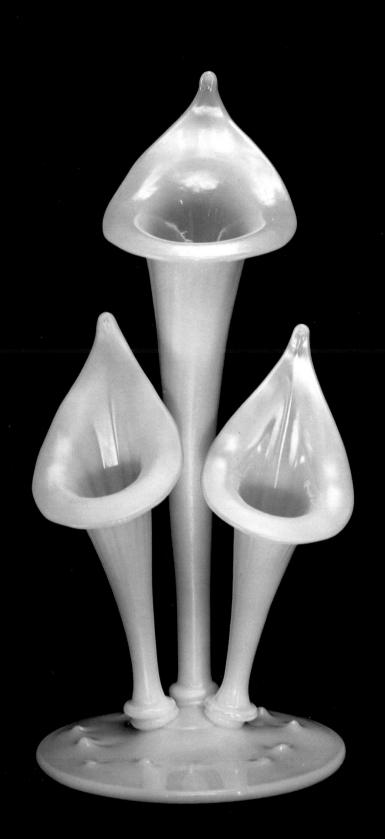

54

53   Free-form bowl, Ivrene. Late 1920s.

54   Wisteria bowl. Mold-blown with polished
     rim. This dichroic glass, called Alexan-
     drite by European glassmakers, is bluish in
     daylight and pinkish in incandescent light.
     1920s.

55

55   Crystal glass slippers. A special order for a theatrical production, "A Kiss for Cinderella," according to Carder but impossible to verify. About 1929.

56   Black glass group. Gold Aurene stoppers in toilet bottles; vase etched in Medieval pattern. 1920s.

57

57 Alabaster vases with applied decoration in
    Black glass, probably 1925.

58 Lamp. Shades are purple cased over
    Alabaster with acid-etched designs. Three
    bronze figures on Alabaster base with
    acid-etched design, 1920s.

60

59   Prong vase. Jade Green on Alabaster base,
     about 1930.

60   Intarsia group. Late 1920s, early 1930s.

62

62 *Pâte de verre* Dancing Fawn panel. Designed and produced by Frederick Carder, 1930s.

THE
STUDIO PERIOD
1932·1963

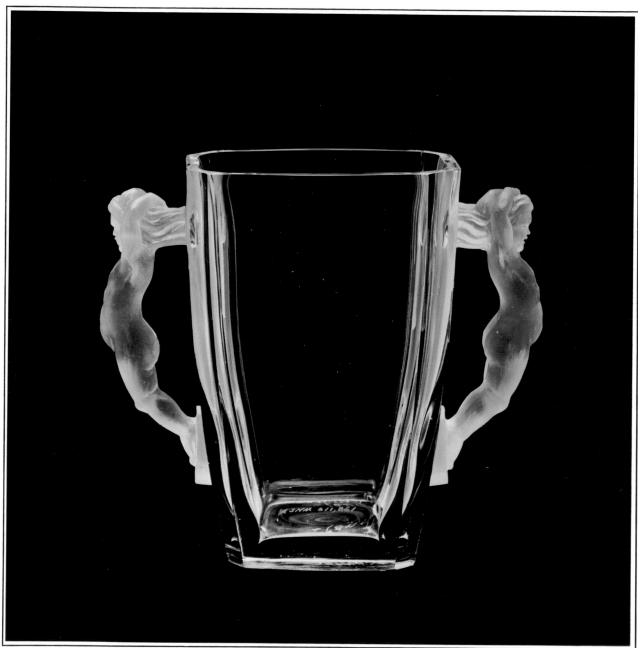

64

63 Female head with scarf. Lost wax casting
in Crystal with satin finish; designed,
modeled, and cast by Frederick Carder,
1930s.

64 Vase with cast handles, blown center. The
cast *cire perdue* (lost wax) handles of
sculptured putti were reheated and fused
to the blown glass vase in the blowing
room. After annealing, the center portion
was ground and polished; 1930s.

65

65   Crystal vase with black flaking. An early
      lost wax casting made before the ceramic
      mold formula was perfected. Black flecks
      from the mold materials combined with
      the melting glass elements to give the verti-
      cal striping effect. Outer surface ground
      and polished, satin finish handles. Unique,
      about 1933.

66   Dragon bowl. Lost wax casting in Crystal
      glass, satin finish. Designed and produced
      by Frederick Carder, mid 1930s.

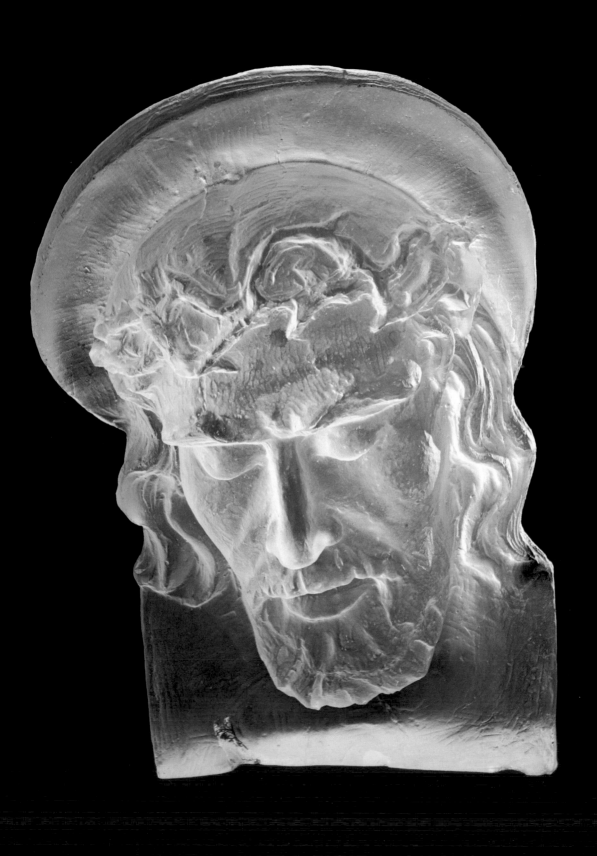

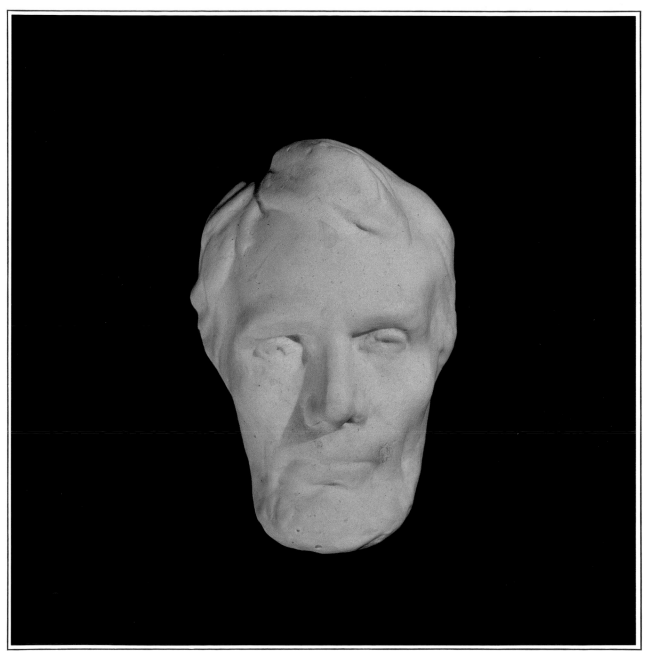

68

67 Head of Christ. Modeled in clay and cast in plaster of Paris by Frederick Carder, October 1933; cast in crystal glass by *cire perdue* process, 1934. Satin finish.

68 Portrait head of Lincoln. Lost wax casting made by Frederick Carder from a Gutzon Borglum model for the Mount Rushmore National Memorial portrait sculpture, about 1937.

69

69 Puma Killing Snake. Lost wax casting in Crystal glass, satin finish. Designed and produced by Frederick Carder. 1941.

70 Figure of a Young Man. Lost wax casting in Crystal glass, satin finish. Designed and produced by Frederick Carder using his grandson, Gillett Welles, as the model. Mid 1930s.

72

71   Diatreta vase. Lost wax casting in Crystal glass, satin finish. Designed and produced by Frederick Carder, probably in the early 1950s.

72   Diatreta vase. Lost wax casting, satin finish. Designed and produced by Frederick Carder, 1953.

74

73 Diatreta vase. Lost wax casting, satin fin-
ish. Designed and produced by Frederick
Carder. Inscription around top "ART IS
LONG/LIFE IS SHORT," 1955.

74 Pilaster capital. Relief in bluish-green glass
cast in an iron mold. Lighted from within;
originally a special order, it was later re-
leased for sale as a stock item, 1930s.

75

75   Self-portrait of Frederick Carder. Lost wax casting in Crystal glass. Modeled and produced by Frederick Carder, 1951.

OBJECT INDEX

24. Nine drinking glasses
   H. (tallest goblet) 20.1 cm. Left row R76.177,* with Steuben fleur-de-lis; R76.201*; R76.203. Center row R76.306,* signed "Carder STEUBEN" (late signature); R76.373.2,* signed "Carder STEUBEN" (late signature); R76.312, acid-stamped "STEUBEN." Right row R78.895*; R78.523, acid-stamped "STEUBEN"; R78.542.

25. Tyrian disk*
   D. 40.4 cm. CMG73.4.85 (Gift of Otto W. Hilbert).

26. Grotesque glass group
   H. (tallest) 45.7 cm. L.-R. R76.569,* signed "F Carder STEUBEN" (late signature); Private collection with Steuben fleur-de-lis; R82.4.339.2 (Bequest of Frank and Mary Elizabeth Reifschlager), with Steuben fleur-de-lis; R76.571 with oval silver paper label "Steuben"; RL83.58.215 acid-stamped "Steuben."

27. Grotesque bowl*
   W. 30.6 cm. Private Collection, stamped with Steuben fleur-de-lis.

28. Grotesque bowl
   H. 12.7 cm. R76.478, signed "STEUBEN."

29. Hexagonal vase*
   H. 30 cm. CMG70.4.104 (Gift of Gillett Welles), signed "F. Carder" (late signature).

30. Vase of Plum Jade*
   H. 29.5 cm. R78.144.

31. Bowl of Black cased over white Cintra, greenish tint*
   H. 15 cm. R79.107, acid-etched Steuben fleur-de-lis.

32. Vase of white cased over Celeste Blue*
   H. 31.9 cm. R82.4.9 (Bequest of Frank and Mary Elizabeth Reifschlager).

33. Moss Agate vase*
   H. 29.7 cm. Collection Dr. and Mrs. L. G. Wagner.

34. Cintra group
   H. (tallest) 39.4 cm. L.-R. CMG75.4.118* (Gift of Corning Glass Works); R77.279, paper label "Steuben"; R82.4.15 (Bequest of Frank and Mary Elizabeth Reifschlager).

35. Cologne bottles
   H. (bottle on left) 19.3 cm. R80.24; R78.563, engraved factory signature "Fred'k Carder"; R78.565,* "F Carder 1918" (late signature).

36. Cluthra vase*
   H. 25 cm. R76.7.

37. Rose Quartz vase*
   H. 28.5 cm. R82.4.143 (Bequest of Frank and Mary Elizabeth Reifschlager).

38. Silverina bowl and candlesticks
   D. (bowl) 28 cm. RL83.58.156, signed "F Carder STEUBEN" (late signature); R76.555.1,2.*

39. Covered jar of Moonlight-colored glass*
   H. 38.2 cm. R76.55, signed "STEUBEN F. Carder" (late signature).

40. Place setting, Wheat pattern*
   D. (plate) 22.2 cm. L.-R. R76.774; R76.772, signed "F Carder STEUBEN"; R76.771, signed "F Carder STEUBEN"; R76.773, signed "F Carder STEUBEN"; bowl and plate R76.770, R78.890, bowl acid-stamped fleur-de-lis, "F Carder STEUBEN" (late signature).

41. Peacock*
   H. 15.7 cm. R82.4.97 (Bequest of Frank and Mary Elizabeth Reifschlager)

42. Vase, Flemish Blue cased over Crystal*
   H. 19.8 cm. R82.4.255 (Bequest of Frank and Mary Elizabeth Reifschlager), acid-stamped "Steuben."

43. Console set, Flemish Blue cased over Rosa (Alexandrite)
   H. (candlesticks) 30.3 cm. Private Collection; bowl with Steuben fleur-de-lis.

44. Console set* and vase, Rose cased over Alabaster
   H. (candlesticks) 36.9 cm. R76.296, with Steuben fleur-de-lis; R76.342.1,2 with Steuben fleur-de-lis; R76.291, gold triangular paper label, "F Carder STEUBEN" (late signature).

45. Goblet. Medallions of Flemish Blue over Crystal, Music pattern*
   H. 22.9 cm. CMG70.4.118 (Gift of Otto W. Hilbert).

46. Stemware
   H. (Black Crystal goblet) 26.1 cm. L.-R. R78.8.15,* acid-stamped STEUBEN; R78.888*; R76.877, acid-stamped STEUBEN, "FC STEUBEN" (late signature); CMG75.4.132* (Gift of Corning Glass Works); R78.9.8, diamond-point "S"; R78.9.66, diamond-point "S"; R78.9.56, diamond-point "Steuben"; R78.9.46 (Gifts of Mr. and Mrs. Freemont Peck), diamond-point "Steuben."

47. Mandarin Yellow group
   H. (tallest) 15.6 cm. L.-R. CMG69.4.238 (Bequest of Gladys C. Welles); CMG55.4.26* (Gift of maker), signed "F Carder 1917"; CMG70.4.109 (Gift of Gillett Welles), signed "F Carder."

48. Rouge Flambé bowl*
   D. 16 cm. R82.4.45 (Bequest of Frank and Mary Elizabeth Reifschlager).

49. Oriental Jade and Oriental Poppy
   H. (vase on left) 17.3 cm. L.-R. R78.539; R82.4.93* (Bequest of Frank and Mary Elizabeth Reifschlager); R78.534*; R78.532; R78.527, signed "F Carder STEUBEN" (late signature).

50. Florentia group
   H. (vase on left) 32.8 cm. L.-R. R77.316,* with Steuben fleur-de-lis; R77.311, with Steuben fleur-de-lis; R77.312.1,2; with Steuben fleur-de-lis; bowl R82.4.316 (Bequest of Frank and Mary Elizabeth Reifschlager), no fleur-de-lis, but blotch remains to receive stamp.

51. Candlesticks and compote
   H. (left) 28.2 cm. L.-R. R82.8.2; R77.31,* with Steuben fleur-de-lis; R82.8.3 (Candlesticks gift of Dr. Martin E. Nordberg).

52. Lily vase*
   H. 30.1 cm. CMG59.4.303 (Gift of maker).

53. Free-form bowl, Ivrene
    W. 36 cm. R77.20.

54. Wisteria bowl*
    H. 12.8 cm. Private Collection, stamped "Steuben."

55. Crystal glass slippers*
    L. 21.4 cm. CMG75.4.166 (Gift of Corning Glass Works); CMG 66.4.74.

56. Black glass group
    H. (tallest) 30.2 cm. L.-R. R76.552.1,2; CMG51.4.704 (Gift of Frederick Carder), triangular gold label with Steuben fleur-de-lis, circular paper labels "STEUBEN GLASS WORKS, CORNING, N.Y."; R77.228* with Steuben fleur-de-lis; R82.4.322* (Bequest of Frank and Mary Elizabeth Reifschlager), with Steuben fleur-de-lis.

57. Alabaster vases with applied decoration in Black glass
    H. (center) 20 cm. L.-R. R78.1131, R83.58.3, R83.58.1.*

58. Lamp with bronze figures
    H. 69 cm. CMG59.4.524 (Gift of Corning Glass Works).

59. Prong vase
    H. 37.7 cm. R76.353, signed "STEUBEN F. Carder" (late signature).

60. Intarsia group
    H. (tallest) 24.7 cm. L.-R. R82.4.89 (Bequest of Frank and Mary Elizabeth Reifschlager), engraved signature "Fred'k Carder"; R83.59.29; CMG69.4.217* (Bequest of Gladys C. Welles), engraved signature "Fred'k Carder, stamped with Steuben fleur-de-lis;

CMG69.4.216 (Bequest of Gladys C. Welles), engraved signature "Fred'k Carder," signed on base "F. Carder" (late signature); bowl R78.814, engraved signature "Fred'k Carder."

61. Intarsia vase
    H. 17.4 cm. CMG69.4.221 (Bequest of Gladys C. Welles). Engraved facsimile signature, "Fred'k Carder."

The Studio Period · 1932-1963

62. *Pâte de verre* Dancing Fawn panel*
    L. 23.6 cm. CMG69.4.255 (Bequest of Gladys C. Welles), signed "Fred'k Carder."

63. Female head with scarf*
    H. 30.9 cm. CMG52.4.332 (Gift of maker). "F.C." monogram on lower left side.

64. Vase with cast handles
    H. 22.3 cm. R78.832, signed "F. Carder 1937."

65. Crystal vase with black flaking*
    H. 16.7 cm. CMG69.4.86 (Gift of Mr. and Mrs. Gillett Welles), signed "F. CARDER."

66. Dragon bowl*
    H. 23 cm. R78.824, signed "STEUBEN F. Carder."

67. Head of Christ
    O.H. 38.7 cm. RL83.58.238.

68. Portrait head of Lincoln*
    H. 11 cm. CMG75.4.152 (Gift of Corning Glass Works).

69. Puma Killing Snake*
    L. 20 cm. CMG75.4.151 (Gift of Corning Glass Works), signed "F. Carder 1941."

70. Figure of a Young Man*
    H. 22.5 cm. CMG52.4.317 (Gift of maker), signed "F. Carder."

71. Diatreta vase*
    H. 19.1 cm. R80.27, signed "F. Carder."

72. Diatreta vase*
    H. 16.2 cm. CMG53.4.26 (Gift of maker), signed "F. Carder 1953."

73. Diatreta vase
    H. 22.9 cm. CMG69.4.251 (Bequest of Gladys C. Welles), signed "F. Carder 1955."

74. Pilaster capital*
    H. 26.8 cm. CMG59.4.525 (Gift of Frederick Carder).

75. Self-portrait of Frederick Carder*
    D. 22.5 cm. CMG59.4.362 (Gift of Corning Glass Works), "FC" monogram cast on center front just above base; "FC" monogram with date, signed on back "F. Carder 1951."

Cover: Cluthra group. H. (vase on left) 30 cm. L.-R. R76.44,* signed "F. Carder STEUBEN"; CMG 72.4.182 (Bequest of Gillett Welles), signed "F. Carder STEUBEN" (both late signatures); R76.6.

Frederick Carder
Selected Bibliography

Arwas, Victor.
Glass: Art Nouveau to Art Deco. New York: Rizzoli, 1977. "Steuben Glassworks," pp. 198-206, ill.; "Stevens & Williams," pp. 207-210, ill.

Bardof, Frank.
"Frederick Carder, Artist in Glass." *Glass Industry,* v. 20, no. 1, April 1939, pp. 136-140.

Buechner, Thomas S.
"The Glass of Frederick Carder." In: *The Connoisseur Year Book, 1961,* [London]: Connoisseur, 1961, pp. 39-43, ill.

Carder, Frederick.
"Art in Glass." In: *Handbook of American Glass Industries,* compiled for the 1936 exhibition of the American Glass Industries, Department of Industrial Art, The Brooklyn Museum, pp. 83-93, ill.

_____.
"Artistic Glass From 1910 to the Present Day." *The Glass Industry,* v. 14, no. 3, March 1934, pp. 28-29, ill.

Carder, Frederick with Frederic Schuler.
The Autobiography of an Englishman in the United States of America. [Corning, N. Y.]: The Corning Museum of Glass, 1957. 22 pp., typed.

Corning Museum of Glass.
Frederick Carder: His Life and Work. Text by Thomas S. Buechner. Corning, N. Y.: The Corning Museum of Glass, 1952. 23 pp., ill.

Ericson, Eric E.
A Guide to Colored Steuben Glass, 1903-1933. Loveland, Colo.: Lithographic Press, [1963-1965]. 2 vol., ill.

Farrar, Estelle Sinclaire and Spillman, Jane Shadel.
The Complete Cut & Engraved Glass of Corning. New York: Crown, 1979. "Steuben Glass Works," pp.195-202, ill.

Gardner, Paul V.
The Glass of Frederick Carder. New York: Crown, [1971]. 373 pp., ill.

Grover, Ray and Lee.
Art Glass Nouveau. Rutland, Vermont: Charles E. Tuttle, 1967, 231 pp., ill.

Haden, Jack.
"Frederick Carder—Designer, Technologist and Centenarian." *Glass Technology,* v. 5, no. 3, June 1964, pp. 105-109, ill.

Hotchkiss, John F.
Carder's Steuben Glass: Handbook and Price Guide.... [2nd ed. Pittsfield, N. Y.: Hotchkiss House, 1972]. 119 pp., ill.

Kamm, Minnie Watson.
"Frederick Carder: Artist in Glass." *Spinning Wheel,* v. 6, no. 10, October 1950, pp. 36-41, ill.

Madigan, Mary Jean.
Steuben Glass: An American Tradition in Crystal. New York: Harry N. Abrams, 1982. 320 pp., ill.

Northwood, John.
John Northwood: His Contributions to the Stourbridge Flint Glass Industry. Stourbridge: Mark and Moody, 1958. 134 pp., ill.

Perrot, Paul N.
"Frederick Carder's Legacy to Glass." *Craft Horizons,* v. 21, no. 3, May/June 1971, pp. 32-35., ill.

Perrot, Paul N.; Gardner, Paul V.; and Plaut, James S.
Steuben: Seventy Years of American Glass Making. [Catalog of an exhibition.] New York: Praeger, [1974]. 172 pp., ill.

Philpot, Gerry.
Creations by Carder of Steuben: His American Art Glass. [Glenbrook, Conn.]: Fieldstone Porch, [1963]. 12 pp. and two L.P. records titled "Conversations with Carder on Steuben...."

Revi, Albert Christian.
American Art Nouveau Glass. Nashville, Tenn.: Thomas Nelson, 1968. "Steuben Glass Works," pp. 128-182, ill.

_____.
"Clutha and Cluthra Glass." *Spinning Wheel,* v. 19, nos. 1-2, January-February 1963, p. 20, ill.

_____.
"Frederick Carder's Patents." *Spinning Wheel,* v. 20, nos. 1-2, January-February, 1964, pp. 14-15.

Rockwell, Robert F.
Frederick Carder and His Steuben Glass, 1903-1933: From the Rockwell Gallery, Corning, N.Y. Text by Jack Lanahan. West Nyack, N.Y.: Dexter Press, 1966. 33 pp., ill.

Silverman, Alexander.
"Frederick Carder, Artist and Glass Technologist." *Bulletin of the American Ceramic Society.* v. 18, no. 9, Sept. 1939, pp. 343-349, ill.

Turner, William Ernest Stephen.
"The Art of Frederick Carder." *Journal of the Society of Glass Technology.* "Proceedings" Section, v. 23, 1939, pp. 41-43, ill.